MATHadazzles

Volume 8

Reasoning Algebraically with Decimals

SUPER STUMPERS!

Authors

Carole Greenes

Mary Cavanagh

Contributors: Grades 9-11 Students

Eve Armstrong, Max Gao, Johnny Guerra, Connor Harney, Bradley Kaufman, Christiana Luna, Nickoli McKenzie, Krish Patel, Nathan Rios, Sanjana Sarkar, Maximus Smith, Nashawn Tadytin, Rebecca Yacoub

Editors

Tanner Wolfram, Senior Editor
Daniel Lee, Assistant Senior Editor
Jason Luc
Yifan Tian
Ping Chuan (Larry) Yong

Cover Design

Mary Cavanagh

Copyright © 2018 Arizona Board of Regents for and on behalf of

Arizona State University (ASU)

PRIME Center

Tempe, Arizona

MATHadazzles Volume 8
Reasoning Algebraically with Decimals
SUPER STUMPERS

MATHadazzle Super Stumper Puzzles will challenge your logical reasoning abilities, your sense of numbers (different types of numbers, their characteristics, and operations with them), and your persistence in solving problems. Once you start, you won't be able to stop UNTIL you successfully solve all of the puzzles! Good Luck. They are tough.

What is an Algebraic MATHadazzle with decimal numbers? An Algebraic MATHadazzle uses variables to represent numbers in equations and inequalities. In this book of puzzles, n is the symbol that represents a number. A MATHadazzle is a 3-by-3 grid with circles at the end of each row and each column. Some grid cells have clues, given as equations or inequalities, that "describe" the decimal numbers that will fill those cells. Numbers in circles at the ends of the rows and at the bottoms of the columns are the row and column sums.

What's Your Job? Based on clues provided in some of the grid cells, you place the decimal numbers $^-2.0$, $^-1.5$, $^-1.0$, $^-0.5$, $^+0.5$, $^+1.0$, $^+1.5$, $^+2.0$ and $^+2.5$ in the nine cells so that the row and column numbers add to the sums in the circles.

Clues:

> **Equations and Inequalities:** An equation has a single solution. An inequality may have one or more solutions.
>
> **Exponent:** A small number to the right and above a base number indicating the number of times that base number should be used as a factor in multiplication. The exponent may be positive or negative.
>
> n^2 means $n \times n$
>> Example: $1.5^2 = 1.5 \times 1.5$, or 2.25
>
> n^{-2} means $1 \div (n \times n)$, or $1 \div n^2$
>> Example: $1.5^{-2} = 1 \div (1.5 \times 1.5)$, or $\frac{1}{2.25}$
>
> n^3 means $n \times n \times n$
>> Example: $2.0^3 = 2.0 \times 2.0 \times 2.0$, or 8.0
>
> n^{-3} means $1 \div (n \times n \times n)$, or $1 \div n^3$
>> Example: $2.0^{-3} = 1 \div (2.0 \times 2.0 \times 2.0)$, or $\frac{1}{8.0}$.

Factorial: This is the product of a number and all counting numbers less than it. The symbol for factorial is the exclamation mark (!).

Example: 4! = 4 x 3 x 2 x 1, or 24.

Operations: All four operations are incorporated into the clues. These are: addition, subtraction, multiplication, and division.

Order of Operations: When several different operations appear in a number sentence, then the order in which they are performed follows the Fundamental Order of Operations: Parentheses (all operations within parentheses are computed first), Powers (all base numbers with exponents), Multiplication and Division (from left to right), and finally, Addition and Subtraction (from left to right).

Proportion: A proportion is an equation showing that two ratios are equivalent. In a proportion, cross products are equal.

Example: $\frac{n-4}{5} = \frac{-0.8-n}{3}$, so 5 x (-0.8-n) = (n-4) x 3

$n = {}^+1.0$

Square Root of a Number: Number that when multiplied by itself has a product equal to a given number. The square root of 4.0 is either $^+2.0$ or $^-2.0$ because: $^+2.0$ x $^+2.0$ = 4.0 and $^-2.0$ x $^-2.0$ = $^+4.0$. The square root symbol is $\sqrt{\ }$.

Example: $\sqrt{9.0} = {}^+3.0$ or $\sqrt{9.0} = {}^-3.0$.

Absolute Value: The absolute value of a number is its positive value regardless of its sign. Absolute value is designated with a pair of parallel line segments surrounding the number.

Example: $|{}^+2.0| = {}^+2.0$ and $|{}^-2.0| = {}^+2.0$

The absolute value of $^+2.0 = {}^+2.0$. The absolute value of $^-2.0 = {}^+2.0$.

If you are interested in improving your dazzling solution talents, consider getting Volumes 1, 2, 3, 4, 5, 6, and 7 with more puzzles. Answers are at the back of each book.

Enjoy Solving!

Your MATHadazzling Authors, Contributors, and Editors

Put these numbers in the squares -2.0, -1.5, -1.0, -0.5, +0.5, +1.0, +1.5, +2.0, +2.5

Add across →

Add down ↓

Sums are in ◯

$\sqrt{n^4 + 9} = -7 + 2$		$1 \div 0.4 = n$
$1 \div 0.5 = n$		$1.5 - n = \dfrac{0}{7}$

Right side circles: **1.5**, **-1.0**, **2.0**

Bottom circles: **-0.5**, **-1.5**, **4.5**

PRIME Center   MATHadazzles Volume 8

2

Put these numbers in the squares -2.0, -1.5, -1.0, -0.5, +0.5, +1.0, +1.5, +2.0, +2.5

Add across →

Add down ↓

Sums are in ◯

$n^7 + n^3 = 2n$		$\dfrac{n}{0.5} + 4 = 0$	**-0.5**
			0.5
$\dfrac{7.5}{n} = n + 0.5$		$n(n - 2.5) = -n$	**2.5**
3.0	**1.0**	**-1.5**	

PRIME Center   MATHadazzles Volume 8

3

Put these numbers in the squares -2.0, -1.5, -1.0, -0.5, +0.5, +1.0, +1.5, +2.0, +2.5

Add across →

Add down ↓

Sums are in ◯

	$-\frac{6n}{9} = n + 2.5$		
			-4.5
		$n = \sqrt[4]{1.0}$	
			5.5
$\frac{2n-3}{6} = -\frac{2}{3}$		$2n = 0.75 \times 4$	
			1.5
0	**1.0**	**1.5**	

PRIME Center   MATHadazzles Volume 8

Put these numbers in the squares -2.0, -1.5, -1.0, -0.5, +0.5, +1.0, +1.5, +2.0, +2.5

Add across →

Add down ↓

Sums are in ◯

$n + 2.5 = 5$			2.5
		$n + 2.5 = 1$	1.0
$n + 6.4 = 7.9$			-1.0
6.0	1.0	-4.5	

PRIME Center

5

Put these numbers in the squares -2.0, -1.5, -1.0, -0.5, +0.5, +1.0, +1.5, +2.0, +2.5

Add across ⟶

Add down ↓

Sums are in ◯

			-3.0
	$n^2 = 62.5 \times 0.1$	$\frac{10}{100} \times 5 = \frac{1}{4}n$	2.5
$\frac{12}{8} = n$			3.0
-1.0	2.5	1.0	

PRIME Center   MATHadazzles Volume 8

6

Put these numbers in the squares -2.0, -1.5, -1.0, -0.5, +0.5, +1.0, +1.5, +2.0, +2.5

Add across ⟶

Add down ↓

Sums are in ◯

$n^3 = 15.625$		$0! \times 1.5 = n$	4.5
			-4.5
		$12n^2 \times \sqrt{9} = 12^2$	2.5
2.5	-2.0	2.0	

PRIME Center MATHadazzles Volume 8

Put these numbers in the squares -2.0, -1.5, -1.0, -0.5, +0.5, +1.0, +1.5, +2.0, +2.5

Add across →

Add down ↓

Sums are in ◯

$\sqrt[5]{-32} = n$			**-4.5**
	$\dfrac{n+2}{n-1.5} = \dfrac{9}{2}$		**4.0**
$\dfrac{6-n}{n+2} = \dfrac{9}{7}$		$0 < n < 1$	**3.0**
1.5	**2.0**	**-1.0**	

PRIME Center MATHadazzles Volume 8

8

Put these numbers in the squares -2.0, -1.5, -1.0, -0.5, +0.5, +1.0, +1.5, +2.0, +2.5

Add across →

Add down ↓

Sums are in ◯

$\frac{4\times 2}{4} + n = 0$		$\frac{4}{3} \times 0 + n = -1.5$	-1.5
			4.5
		$\frac{5-3}{2} + n = 0.5$	-0.5
-2.5	5.5	-0.5	

PRIME Center

MAThadazzles Volume 8

9

Put these numbers in the squares -2.0, -1.5, -1.0, -0.5, +0.5, +1.0, +1.5, +2.0, +2.5

Add across →

Add down ↓

Sums are in ◯

		$0.3 = 0.6n$	 -1.0
			 -1.5
$0.5 + 1 + n = 3$		$\dfrac{5n+3}{3+n} = 2$	 5.0
◯ -1.5	◯ 0.5	◯ 3.5	

PRIME Center   MATHadazzles Volume 8

10

Put these numbers in the squares -2.0, -1.5, -1.0, -0.5, +0.5, +1.0, +1.5, +2.0, +2.5

Add across →

Add down ↓

Sums are in ◯

$3^3 = n + 26$		
	$n^2 = 6.25$	
$n = n \times 2 - 2$		$5 - 3 - 1.5 = n$

Right column sums: 2.0, -0.5, 1.0

Bottom row sums: 1.0, 0.5, 1.0

PRIME Center MATHadazzles Volume 8

11

Put these numbers in the squares   -2.0, -1.5, -1.0, -0.5, +0.5, +1.0, +1.5, +2.0, +2.5

Add across ⟶

Add down ↓

Sums are in ◯

			2.5
$2n + 6 = 3$			-0.5
$\dfrac{\left(\frac{4}{6}\right)}{n-1} = -\dfrac{4}{18}$	$\dfrac{0.4}{2} = \dfrac{0.8n}{2}$		0.5
-4.5	2.5	4.5	

PRIME Center   MATHadazzles Volume 8

Put these numbers in the squares -2.0, -1.5, -1.0, -0.5, +0.5, +1.0, +1.5, +2.0, +2.5

Add across ⟶

Add down ↓

Sums are in ◯

	$\sqrt[3]{4n} \times n = n^2$		3.5
$n^3 = -3.375$		$\dfrac{4}{1.6} = n$	0.5
			-1.5
-3.0	3.0	2.5	

PRIME Center

13

Put these numbers in the squares -2.0, -1.5, -1.0, -0.5, +0.5, +1.0, +1.5, +2.0, +2.5

Add across →

Add down ↓

Sums are in ◯

$\frac{1}{3}n = -\frac{1}{3}$		$\frac{n^2}{16} = \frac{1}{8}n$
	$0.5n = n - 0.5$	

Right circles: **-0.5**, **4.0**, **-1.0**

Bottom circles: **1.0**, **1.0**, **0.5**

PRIME Center MATHadazzles Volume 8

14

Put these numbers in the squares -2.0, -1.5, -1.0, -0.5, +0.5, +1.0, +1.5, +2.0, +2.5

Add across →

Add down ↓

Sums are in ◯

			-4.5
	$\dfrac{50n}{8} = n^3$	$n^3 = 0.125$	**2.5**
		$\dfrac{n^3 - 6}{2} = 1$	**4.5**
-1.0	**2.5**	**1.0**	

PRIME Center　　　　　　　　　　　MATHadazzles Volume 8

15

Put these numbers in the squares -2.0, -1.5, -1.0, -0.5, +0.5, +1.0, +1.5, +2.0, +2.5

Add across →

Add down ↓

Sums are in ◯

		$\frac{2n-3}{6} = -\frac{2}{3}$	**-0.5**
			3.0
$5n^3 = -5$		$\frac{n+0.5}{2} = \frac{1}{-2}$	**0**
-1.5	**5.0**	**-1.0**	

PRIME Center

MATHadazzles Volume 8

16

Put these numbers in the squares -2.0, -1.5, -1.0, -0.5, +0.5, +1.0, +1.5, +2.0, +2.5

Add across →

Add down ↓

Sums are in ◯

$2n^2 + n - 1 = 0$		$n > 0$
	$n^2 = -\dfrac{n}{2}$	$4n^2 - 25 = 0$

Right column circles: -0.5, -1.0, 4.0

Bottom row circles: 1.0, -1.0, 2.5

PRIME Center   MATHadazzles Volume 8

17

Put these numbers in the squares -2.0, -1.5, -1.0, -0.5, +0.5, +1.0, +1.5, +2.0, +2.5

Add across →

Add down ↓

Sums are in ◯

	$\frac{6}{8}n = -\frac{3}{4}$		
			-0.5
	$-\frac{2}{3}(2.5n) = \frac{10}{3}$		
			2.5
		$0.5 \times 2^0 = n$	
			0.5
3.5	-4.5	3.5	

PRIME Center   MATHadazzles Volume 8

18

Put these numbers in the squares -2.0, -1.5, -1.0, -0.5, +0.5, +1.0, +1.5, +2.0, +2.5

Add across ⟶

Add down ↓

Sums are in ◯

$n + \dfrac{236}{2} = 117.5$	$\dfrac{n+258}{26} = \dfrac{257}{26}$	
	$\sqrt[4]{16} - n = 0.5$	

Right side circles: 0.5, -3.0, 5.0

Bottom circles: 1.0, -1.0, 2.5

PRIME Center

19

Put these numbers in the squares -2.0, -1.5, -1.0, -0.5, +0.5, +1.0, +1.5, +2.0, +2.5

Add across →

Add down ↓

Sums are in ◯

$0.16 = 0.32n$			 -2.0
		$\dfrac{2}{n} = n - 1$	 1.5
$2n^0 = n + n$		$2n^2 = -n$	 3.0
-0.5	3.0	0	

PRIME Center MATHadazzles Volume 8

20

Put these numbers in the squares -2.0, -1.5, -1.0, -0.5, +0.5, +1.0, +1.5, +2.0, +2.5

Add across →

Add down ↓

Sums are in ◯

$n < 0$	$n^3 + 4n^2 - 4 = n$	
$8n = 20$		$\dfrac{1}{n} \times 5 = 10$

Row sums: -2.0, 4.5, 0

Column sums: 2.5, 1.0, -1.0

21

Put these numbers in the squares -2.0, -1.5, -1.0, -0.5, +0.5, +1.0, +1.5, +2.0, +2.5

Add across →

Add down ↓

Sums are in ◯

$(n-2)^0 = n$		$n^2 = \dfrac{25}{4}$	**4.0**
			-2.0
		$2^n = 4$	**0.5**
-1.5	**1.0**	**3.0**	

PRIME Center

22

Put these numbers in the squares -2.0, -1.5, -1.0, -0.5, +0.5, +1.0, +1.5, +2.0, +2.5

Add across →

Add down ↓

Sums are in ◯

$4n = \sqrt{100}$		$\dfrac{4 \times 2.5}{5} \times n = 4$
	$2n + 2 = 0$	

Row sums: (2.5), (-1.5), (1.5)

Column sums: (3.0), (-2.5), (2.0)

PRIME Center   MATHadazzles Volume 8

23

Put these numbers in the squares -2.0, -1.5, -1.0, -0.5, +0.5, +1.0, +1.5, +2.0, +2.5

Add across →

Add down ↓

Sums are in ◯

		$n = 0.5 \times 2 - 0.5$	-0.5
			-0.5
$n = 4 \times \dfrac{10}{25} + 0.4$		$n = 4 \times 0.5 + 0.5$	3.5
4.5	-3.5	1.5	

PRIME Center MATHadazzles Volume 8

24

Put these numbers in the squares   -2.0, -1.5, -1.0, -0.5, +0.5, +1.0, +1.5, +2.0, +2.5

Add across ⟶

Add down ↓

Sums are in ◯

$n^2 + \dfrac{n}{2} = 2n$			0
		$4n \div 3 = n - 0.5$	-3.0
$2 \div 0.8 = n$			5.5
3.0	-0.5	0	

PRIME Center MAThadazzles Volume 8

25

Put these numbers in the squares -2.0, -1.5, -1.0, -0.5, +0.5, +1.0, +1.5, +2.0, +2.5

Add across →

Add down ↓

Sums are in ◯

		$\dfrac{n^0}{n} = 1$	**1.0**
$n^{17} = n$			**-4.5**
$n^2 = 2n - 0.75$			**6.0**
1.0	**0.5**	**1.0**	

PRIME Center · MATHadazzles Volume 8

26

Put these numbers in the squares -2.0, -1.5, -1.0, -0.5, +0.5, +1.0, +1.5, +2.0, +2.5

Add across →

Add down ↓

Sums are in ◯

$n^2 + 4n = -4$			**-3.5**
			6.0
$\dfrac{n}{2} = \dfrac{1}{\sqrt[3]{64}}$	$(n+2)^2 = \dfrac{1}{4}$		**0**
0	**0.5**	**2.0**	

Put these numbers in the squares -2.0, -1.5, -1.0, -0.5, +0.5, +1.0, +1.5, +2.0, +2.5

Add across →

Add down ↓

Sums are in ◯

	$\dfrac{3}{2} = \dfrac{2n+1}{3n-1}$	$n^2 = -3n - 2.25$	1.0
			-1.0
		$n^2 + 3.75 = 4n$	2.5
3.0	0.5	-1.0	

PRIME Center   MATHadazzles Volume 8

28

Put these numbers in the squares -2.0, -1.5, -1.0, -0.5, +0.5, +1.0, +1.5, +2.0, +2.5

Add across →

Add down ↓

Sums are in ◯

	$n^3 - 6n = -2n$	
$n > 0$	$n^2 \times 2 = 12.5$	

Row sums: 0, 1.0, 1.5

Column sums: 1.0, 6.0, -4.5

PRIME Center   MATHadazzles Volume 8

Put these numbers in the squares -2.0, -1.5, -1.0, -0.5, +0.5, +1.0, +1.5, +2.0, +2.5

Add across →

Add down ↓

Sums are in ◯

			-4.5
	$n > 0$	$\sqrt{64n} = 8$	1.0
$9 = 4n^2$		$\dfrac{2.5 + n}{4} = \dfrac{4n}{8}$	6.0
-1.0	1.0	2.5	

30

Put these numbers in the squares -2.0, -1.5, -1.0, -0.5, +0.5, +1.0, +1.5, +2.0, +2.5

Add across →

Add down ↓

Sums are in ◯

		$(n + 0.5)! = 6$	**-1.0**
$-1 = \left(\dfrac{7}{9} + \dfrac{6}{27}\right)n$		$\dfrac{n + 4.5}{2(n + 2.5)} = 1$	**0**
	$16^{\frac{1}{4}} \div 4 = n$		**3.5**
-0.5	**0**	**3.0**	

PRIME Center   MATHadazzles Volume 8

31

Put these numbers in the squares -2.0, -1.5, -1.0, -0.5, +0.5, +1.0, +1.5, +2.0, +2.5

Add across →

Add down ↓

Sums are in ◯

		$\dfrac{\sqrt{n} \times 0.5}{\frac{1}{4}} = 2$	(5.5)
			(-3.0)
	$\dfrac{n+1}{2} = \dfrac{1}{4}$	$n = \dfrac{-1000^0 \times n}{n}$	(0)
(2.5)	(2.0)	(-2.0)	

PRIME Center   MATHadazzles Volume 8

Put these numbers in the squares -2.0, -1.5, -1.0, -0.5, +0.5, +1.0, +1.5, +2.0, +2.5

Add across →

Add down ↓

Sums are in ◯

		$n^2 = -n$	-4.5
		$3^n = \dfrac{\sqrt{81}}{2} - 1.5$	1.0
	$4^n = 2\sqrt{64}$		6.0
-1.0	1.0	2.5	

33

Put these numbers in the squares -2.0, -1.5, -1.0, -0.5, +0.5, +1.0, +1.5, +2.0, +2.5

Add across →

Add down ↓

Sums are in ◯

			-1.0
$(n+0.5)^5 = 3^{(2n)}$		$n^4 = 4^n$	2.5
	$2n = \sqrt[3]{27}$		1.0
1.0	0.5	1.0	

PRIME Center

34

Put these numbers in the squares -2.0, -1.5, -1.0, -0.5, +0.5, +1.0, +1.5, +2.0, +2.5

Add across →

Add down ↓

Sums are in ◯

		$5.0 = 11.2n - 0.6$	**0**
	$2n + 15 = 18$		**1.5**
		$26 = n + 3^3$	**1.0**
2.5	**2.5**	**-2.5**	

PRIME Center　　　　　　　　　　　　　MATHadazzles Volume 8

Put these numbers in the squares -2.0, -1.5, -1.0, -0.5, +0.5, +1.0, +1.5, +2.0, +2.5

Add across →

Add down ↓

Sums are in ◯

		$16n = 32n^2$	**0**
	$\sqrt{n+7.5} = 2n$		**1.5**
		$4! = n + 25$	**1.0**
1.5	**-0.5**	**1.5**	

Put these numbers in the squares -2.0, -1.5, -1.0, -0.5, +0.5, +1.0, +1.5, +2.0, +2.5

Add across →

Add down ↓

Sums are in ◯

$10^n \times \dfrac{n}{2} = 5$			2.5
		$0.6n = n + 0.2$	0
$10n + 6n = 8$			0
0	-1.0	3.5	

37

Put these numbers in the square -2.0, -1.5, -1.0, -0.5, +0.5, +1.0, +1.5, +2.0, +2.5

Add across →

Add down ↓

Sums are in ◯

	$\frac{(3n)^2}{2} = 10.125$	
$|n^8 - 31n| = 30$		$\sqrt[3]{-0.125} = n$

Row sums: -2.0, 1.0, 3.5

Column sums: -2.0, 4.5, 0

PRIME Center   MATHadazzles Volume 8

Put these numbers in the squares -2.0, -1.5, -1.0, -0.5, +0.5, +1.0, +1.5, +2.0, +2.5

Add across →

Add down ↓

Sums are in ◯

	$8\sqrt[5]{32} = \dfrac{40n}{5}$		2.5
			-2.0
$15.5 = 7n + 5$		$25 = 4n^2$	2.0
-0.5	-1.0	4.0	

39

Put these numbers in the squares -2.0, -1.5, -1.0, -0.5, +0.5, +1.0, +1.5, +2.0, +2.5

Add across →

Add down ↓

Sums are in ◯

	$4n^2 - n = n$		
			0.5
		$\dfrac{2n+6}{2n-5} = -\dfrac{3}{8}$	-1.0
		$\dfrac{3-n}{n} = \dfrac{2n}{3}$	3.0
-3.5	5.0	1.0	

PRIME Center

MATHadazzles Volume 8

40

Put these numbers in the squares -2.0, -1.5, -1.0, -0.5, +0.5, +1.0, +1.5, +2.0, +2.5

Add across →

Add down ↓

Sums are in ◯

$\frac{2}{n} = \sqrt[3]{64}$		$3.2 = \frac{n}{n} \times \frac{8}{n}$	**4.5**
			2.0
	$-6n - 2 = 7$		**-4.0**
-1.0	**1.0**	**2.5**	

PRIME Center   MAThadazzles Volume 8

Put these numbers in the squares -2.0, -1.5, -1.0, -0.5, +0.5, +1.0, +1.5, +2.0, +2.5

Add across ⟶

Add down ↓

Sums are in ◯

	$n^2 - 3n + 2 = 0$	$n^3 = -0.125$	
			-0.5
			-1.5
	$4n = 6$		4.5

-1.0   2.5   1.0

42

Put these numbers in the squares -2.0, -1.5, -1.0, -0.5, +0.5, +1.0, +1.5, +2.0, +2.5

Add across →

Add down ↓

Sums are in ◯

$n^2 = 6.25$			**1.0**
			-1.0
$-1 < n < 0.5$		$\dfrac{6^2 \times 0.5}{9} = n$	**2.5**
0.5	**-2.0**	**4.0**	

43

Put these numbers in the squares -2.0, -1.5, -1.0, -0.5, +0.5, +1.0, +1.5, +2.0, +2.5

Add across ⟶

Add down ↓

Sums are in ◯

			2.0
$\sqrt{6n+6}=3$	$-10=2n-8$		-2.5
		$\sqrt{5.25-3}=n$, $n>0$	3.0
-1.5	2.0	2.0	

PRIME Center

44

Put these numbers in the squares -2.0, -1.5, -1.0, -0.5, +0.5, +1.0, +1.5, +2.0, +2.5

Add across →

Add down ↓

Sums are in ◯

		$n^2 - 3.25 = 3$	**0**
			1.5
$n^{2n} = 3.375$	$n^{\frac{1}{2}} + n^2 = 2$		**1.0**
1.5	**-0.5**	**1.5**	

PRIME Center MATHadazzles Volume 8

45

Put these numbers in the squares -2.0, -1.5, -1.0, -0.5, +0.5, +1.0, +1.5, +2.0, +2.5

Add across →

Add down ↓

Sums are in ◯

		$n^3 - n^2 = -2$	**-4.5**
		$n^8 + n^3 = 2$	**1.0**
$4n = n^2 + 2.5n$			**6.0**
-1.0	**1.0**	**2.5**	

PRIME Center

MATHadazzles Volume 8

Put these numbers in the squares -2.0, -1.5, -1.0, -0.5, +0.5, +1.0, +1.5, +2.0, +2.5

Add across →

Add down ↓

Sums are in ◯

$\frac{1}{n} = n^2$			0.5
	$\sqrt{10n^2 + 1.5} = 8$		0.5
$2.5n = \sqrt{\frac{100}{n^2}}$		$-1 > n$	1.5
1.5	4.5	-3.5	

PRIME Center

47

Put these numbers in the squares -2.0, -1.5, -1.0, -0.5, +0.5, +1.0, +1.5, +2.0, +2.5

Add across ⟶

Add down ↓

Sums are in ◯

	$4n + 4 = 8n^2$		-2.5
	$4! \div n^2 = 3!$		6.0
$3! - 7 = n$			-1.0
-0.5	3.5	-0.5	

PRIME Center

48

Put these numbers in the squares -2.0, -1.5, -1.0, -0.5, +0.5, +1.0, +1.5, +2.0, +2.5

Add across ⟶

Add down ↓

Sums are in ◯

$n - \dfrac{9}{3} + \dfrac{24}{12} = 0$			**2.0**
			-2.0
	$0.2 = 0.4n$	$-0.4 = 0.8n$	**2.5**
2.0	**4.0**	**-3.5**	

PRIME Center MATHadazzles Volume 8

49

Put these numbers in the squares -2.0, -1.5, -1.0, -0.5, +0.5, +1.0, +1.5, +2.0, +2.5

Add across →

Add down ↓

Sums are in ◯

$0 < n < 2.5$	$1 + n = n^3 + 1$	$(n^2 - 1)^2 = 9$	
			1.5
		$\dfrac{2n}{3} = \dfrac{3}{2n}$	**0.5**
			0.5
-3.0	**-0.5**	**6.0**	

PRIME Center MATHadazzles Volume 8

50

Put these numbers in the squares -2.0, -1.5, -1.0, -0.5, +0.5, +1.0, +1.5, +2.0, +2.5

Add across →

Add down ↓

Sums are in ◯

	$4n = 3!$	$\dfrac{2n+1}{n+2} = -\dfrac{1}{n+2}$	⬤ 2.5
$-5 + 4n = 2n$	$-2 < n < -0.5$		⬤ 0.5
			⬤ -0.5
⬤ 5.0	⬤ 1.0	⬤ -3.5	

PRIME Center   MATHadazzles Volume 8

51

Put these numbers in the squares -2.0, -1.5, -1.0, -0.5, +0.5, +1.0, +1.5, +2.0, +2.5

Add across →

Add down ↓

Sums are in ◯

$n^3 \times \sqrt{9} = 24$		$2 - 3.5 = n$	**3.0**
			-1.0
$\dfrac{127.5}{n} = -127.5$			**0.5**
2.5	**1.5**	**-1.5**	

PRIME Center   MATHadazzles Volume 8

Put these numbers in the squares -2.0, -1.5, -1.0, -0.5, +0.5, +1.0, +1.5, +2.0, +2.5

Add across →

Add down ↓

Sums are in ◯

$n^2 + n = 2$, $n < 0$		$6n \div \left(\frac{16}{2}\right) = -\frac{3}{4}$	-1.5
			2.0
$n < 0$		$n = \sqrt[3]{0.125}$	2.0
-1.5	2.0	2.0	

Put these numbers in the squares -2.0, -1.5, -1.0, -0.5, +0.5, +1.0, +1.5, +2.0, +2.5

Add across ⟶

Add down ↓

Sums are in ◯

			2.0
	$n^3 = -3.375$		1.0
$n = 0.1 \times \sqrt{625}$		$2! \div n = -2$	-0.5
4.5	-4.0	2.0	

54

Put these numbers in the squares -2.0, -1.5, -1.0, -0.5, +0.5, +1.0, +1.5, +2.0, +2.5

Add across →

Add down ↓

Sums are in ◯

$n^2 - n + 2.5 = n^2$			6.0
$\dfrac{1}{n^3} = n^5$	$n = \sqrt{\dfrac{n}{2}}$		1.0
		$n^2 + 2n = 0$	-4.5
2.5	1.0	-1.0	

PRIME Center MATHadazzles Volume 8

Put these numbers in the squares -2.0, -1.5, -1.0, -0.5, +0.5, +1.0, +1.5, +2.0, +2.5

Add across →

Add down ↓

Sums are in ◯

	$n^3 + n^2 = 6n$	
$\dfrac{1}{n} = n^2$		$n + 0.5 = \dfrac{n + 0.5}{n - 0.5}$

Row sums: 2.5, 1.5, -1.5

Column sums: 2.0, 1.5, -1.0

56

Put these numbers in the squares -2.0, -1.5, -1.0, -0.5, +0.5, +1.0, +1.5, +2.0, +2.5

Add across →

Add down ↓

Sums are in ◯

		$4n^2 + 4n = 3$	0.5
	$4n^3 - 6n = 10n$ $n > 0$		2.0
		$\dfrac{1}{n} = -2$	0
0.5	5.0	-3.0	

PRIME Center   MATHadazzles Volume 8

57

Put these numbers in the squares -2.0, -1.5, -1.0, -0.5, +0.5, +1.0, +1.5, +2.0, +2.5

Add across →

Add down ↓

Sums are in ◯

		$\frac{n}{5} = \sqrt{n^2 - 6}$	2.5
	$2n + 3 = 0$	$n > 1$	2.0
$\frac{4n-3}{3n-4} = \frac{2n-1}{n-2}$			-2.0
3.0	-3.0	2.5	

PRIME Center

58

Put these numbers in the squares -2.0, -1.5, -1.0, -0.5, +0.5, +1.0, +1.5, +2.0, +2.5

Add across →

Add down ↓

Sums are in ◯

		$n^2 = 1$ $n < 0$	**-0.5**
			0
$n \times n = 6.25$		$n = \|48 - 50\|$	**3.0**
3.5	**-2.5**	**1.5**	

PRIME Center   MATHadazzles Volume 8

Put these numbers in the squares -2.0, -1.5, -1.0, -0.5, +0.5, +1.0, +1.5, +2.0, +2.5

Add across →

Add down ↓

Sums are in ◯

			-1.0
		$\dfrac{2n+2}{10} = \dfrac{3-n}{5}$	2.0
$(8+2n)^2 = 36$	$2 - \dfrac{1}{4}n = 1.5$		1.5
-4.5	4.0	3.0	

PRIME Center   MATHadazzles Volume 8

60

Put these numbers in the squares -2.0, -1.5, -1.0, -0.5, +0.5, +1.0, +1.5, +2.0, +2.5

Add across →

Add down ↓

Sums are in ◯

		$\frac{650}{5} \div n = 130$	**1.5**
	$-4n = 8$		**0**
$\frac{4n^2}{3} = 3$			**1.0**
6.0	**-3.0**	**-0.5**	

PRIME Center

61

Put these numbers in the squares -2.0, -1.5, -1.0, -0.5, +0.5, +1.0, +1.5, +2.0, +2.5

Add across →

Add down ↓

Sums are in ◯

$n^3 = -\dfrac{4!}{3}$			**0.5**
	$n < 0$	$-3n = 4.5$	**0**
	$-\dfrac{3n}{\sqrt{9}} = 1$		**2.0**
0.5	**0**	**2.0**	

PRIME Center MATHadazzles Volume 8

Put these numbers in the squares -2.0, -1.5, -1.0, -0.5, +0.5, +1.0, +1.5, +2.0, +2.5

Add across →

Add down ↓

Sums are in ◯

$n^4 - 20 = 2n$		$\dfrac{5!}{n} = 60$	**-0.5**
			2.5
	$n^4 = 5.0625$ $n < 1$		**0.5**
-0.5	**0.5**	**2.5**	

63

Put these numbers in the squares -2.0, -1.5, -1.0, -0.5, +0.5, +1.0, +1.5, +2.0, +2.5

Add across ⟶

Add down ↓

Sums are in ◯

	$\|3n\| \div 3 = n$ $n < 1$		-2.0
$8n = \sqrt{144}$		$n! = \dfrac{4}{n}$	2.0
		$n^4 \times 32 = 2$	2.5
-1.0	0	3.5	

PRIME Center MATHadazzles Volume 8

64

Put these numbers in the squares -2.0, -1.5, -1.0, -0.5, +0.5, +1.0, +1.5, +2.0, +2.5

Add across →

Add down ↓

Sums are in ◯

$n^2 + n - 0.75 = 3$			6.0
	$\dfrac{n-4}{5} = \dfrac{-0.8-n}{3}$		-1.5
$n \times 0^0 = -1$			-2.0
-1.5	3.5	0.5	

PRIME Center

65

Put these numbers in the squares -2.0, -1.5, -1.0, -0.5, +0.5, +1.0, +1.5, +2.0, +2.5

Add across ⟶

Add down ↓

Sums are in ◯

$\sqrt{8n} = n^2$			**1.0**
		$\sqrt[5]{-32} = n$	**-1.0**
	$n^3 + n^2 + 2n = -2$		**2.5**
4.5	**-1.0**	**-1.0**	

PRIME Center MATHadazzles Volume 8

Put these numbers in the squares -2.0, -1.5, -1.0, -0.5, +0.5, +1.0, +1.5, +2.0, +2.5

Add across ⟶

Add down ↓

Sums are in ◯

$5n = 28 - 38$			
	$2518n = 1259$		**1.0**
	$n > 0$	$2n^2 - n = 10$	**2.5**

Row sum circles (right side, top to bottom): **-1.0**, **1.0**, **2.5**

Column sum circles (bottom, left to right): **-1.0**, **1.0**, **2.5**

Put these numbers in the squares -2.0, -1.5, -1.0, -0.5, +0.5, +1.0, +1.5, +2.0, +2.5

Add across →

Add down ↓

Sums are in ◯

$\frac{(n^2+n)}{3}=1.25$		$\frac{6n^2}{9}=\frac{-12n}{18}$	**2.5**
			0
$\frac{-n}{4}=\frac{n+3}{2}$			**0**
0.5	**0**	**2.0**	

PRIME Center   MATHadazzles Volume 8

68

Put these numbers in the squares -2.0, -1.5, -1.0, -0.5, +0.5, +1.0, +1.5, +2.0, +2.5

Add across →

Add down ↓

Sums are in ◯

	$\frac{5}{6}n = -\frac{5}{12}$	
		$\frac{3}{2}n = 3$
$\frac{12}{50}n = -\frac{6}{25}$		

Row sums: -1.0, 1.0, 2.5

Column sums: -4.5, 1.0, 6.0

PRIME Center MATHadazzles Volume 8

Put these numbers in the squares -2.0, -1.5, -1.0, -0.5, +0.5, +1.0, +1.5, +2.0, +2.5

Add across →

Add down ↓

Sums are in ◯

	$n + \sqrt{16} = 3n$		**3.5**
	$0 < 2n + 5 < 2$	$-\dfrac{24}{10} = \dfrac{17n+5}{5}$	**-4.5**
$n^2 + 2n = -0.75$			**3.5**
-1.0	**2.5**	**1.0**	

PRIME Center

MATHadazzles Volume 8

70

Put these numbers in the squares -2.0, -1.5, -1.0, -0.5, +0.5, +1.0, +1.5, +2.0, +2.5

Add across →

Add down ↓

Sums are in ◯

	$5^2 = \dfrac{50n}{5}$	
		$0.5 < n < 2$
		$\dfrac{4.5}{2} = n^2$ $n > 0$

Row sums: 4.0, 0, -1.5

Column sums: 1.5, -1.0, 2.0

PRIME Center

MATHadazzles Volume 8

71

Put these numbers in the squares -2.0, -1.5, -1.0, -0.5, +0.5, +1.0, +1.5, +2.0, +2.5

Add across ⟶

Add down ↓

Sums are in ◯

		$\dfrac{58+4n}{36-168} = -0.5$	**-1.5**
$n > 0.5$	$\dfrac{n+1.5}{6} = 0.5$		**3.0**
$\dfrac{n+3.5}{n-0.5} = 3$			**1.0**
1.5	**-1.0**	**2.0**	

PRIME Center MATHadazzles Volume 8

Put these numbers in the squares -2.0, -1.5, -1.0, -0.5, +0.5, +1.0, +1.5, +2.0, +2.5

Add across →

Add down ↓

Sums are in ◯

		$2^2 \times \dfrac{0.25}{2} = n$	**-1.0**
		$n > 2$	**2.5**
$4n^{12} = \sqrt{16}$			**1.0**
-1.0	**2.5**	**1.0**	

73

Put these numbers in the squares -2.0, -1.5, -1.0, -0.5, +0.5, +1.0, +1.5, +2.0, +2.5

Add across ⟶

Add down ↓

Sums are in ◯

$n < 2.5$	$\sqrt[3]{-8} = n$	
$(n-1)! = n$		$2n^2 = 2n - 0.5$

(**2.0**) ( ) (**1.5**)

PRIME Center   MAThadazzles Volume 8

74

Put these numbers in the squares -2.0, -1.5, -1.0, -0.5, +0.5, +1.0, +1.5, +2.0, +2.5

Add across →

Add down ↓

Sums are in ◯

			-1.5
	$n = \dfrac{5}{2}$	$n > 1.5$	**2.5**
		$\dfrac{\sqrt[3]{-125}}{10} = n$	**1.5**
-2.5	**2.5**	**2.5**	

75

Put these numbers in the squares -2.0, -1.5, -1.0, -0.5, +0.5, +1.0, +1.5, +2.0, +2.5

Add across →

Add down ↓

Sums are in ◯

$\dfrac{8^2 \times 5}{2n} = 64$			-0.5
		$-1 < n < 1$	2.0
		$n^6 = n$	1.0
0.5	2.5	-0.5	

PRIME Center   MATHadazzles Volume 8

76

Put these numbers in the squares -2.0, -1.5, -1.0, -0.5, +0.5, +1.0, +1.5, +2.0, +2.5

Add across →

Add down ↓

Sums are in ◯

$\dfrac{5}{2n+2.5}=\dfrac{10}{7}$		
	$2n + 4 = 3.6n$	
	$n > 1$	$1 < n < 2$

Right circles: -0.5, 1.0, 2.0

Bottom circles: -1.5, 2.5, 1.5

PRIME Center   MATHadazzles Volume 8

Put these numbers in the squares -2.0, -1.5, -1.0, -0.5, +0.5, +1.0, +1.5, +2.0, +2.5

Add across →

Add down ↓

Sums are in ◯

		$5^4n - 625 = 0$	**-2.5**
	$n > 1.5$		**4.0**
$n > 1$		$n^2 = 1$	**1.0**
2.5	**-0.5**	**0.5**	

Put these numbers in the squares -2.0, -1.5, -1.0, -0.5, +0.5, +1.0, +1.5, +2.0, +2.5

Add across ⟶

Add down ↓

Sums are in ◯

	$-1 \leq n < 0$	$n^2 = \dfrac{1}{n^2}$ $n > 0$	**1.0**
			-0.5
$3n - 2 = \sqrt{9n-2}$			**2.0**
0.5	**0.5**	**1.5**	

PRIME Center   MATHadazzles Volume 8

Answers

1.

-2.0	1.0	2.5	(1.5)
-0.5	-1.0	0.5	(-1.0)
2.0	-1.5	1.5	(2.0)

(-0.5) (-1.5) (4.5)

2.

1.0	0.5	-2.0	(-0.5)
-0.5	2.0	-1.0	(0.5)
2.5	-1.5	1.5	(2.5)

(3.0) (1.0) (-1.5)

3.

-2.0	-1.5	-1.0	(-4.5)
2.5	2.0	1.0	(5.5)
-0.5	0.5	1.5	(1.5)

(0) (1.0) (1.5)

4.

2.5	1.0	-1.0	(2.5)
2.0	0.5	-1.5	(1.0)
1.5	-0.5	-2.0	(-1.0)

(6.0) (1.0) (-4.5)

5.

-0.5	-1.0	-1.5	(-3.0)
-2.0	2.5	2.0	(2.5)
1.5	1.0	0.5	(3.0)

(-1.0) (2.5) (1.0)

6.

2.5	0.5	1.5	(4.5)
-1.0	-2.0	-1.5	(-4.5)
1.0	-0.5	2.0	(2.5)

(2.5) (-2.0) (2.0)

Answers

7.

-2.0	-1.5	-1.0	(-4.5)
2.0	2.5	-0.5	(4.0)
1.5	1.0	0.5	(3.0)
(1.5)	(2.0)	(-1.0)	

8.

-2.0	2.0	-1.5	(-1.5)
0.5	2.5	1.5	(4.5)
-1.0	1.0	-0.5	(-0.5)
(-2.5)	(5.5)	(-0.5)	

9.

-1.0	-0.5	0.5	(-1.0)
-2.0	-1.5	2.0	(-1.5)
1.5	2.5	1.0	(5.0)
(-1.5)	(0.5)	(3.5)	

10.

1.0	-0.5	1.5	(2.0)
-2.0	2.5	-1.0	(-0.5)
2.0	-1.5	0.5	(1.0)
(1.0)	(0.5)	(1.0)	

11.

-1.0	2.5	1.0	(2.5)
-1.5	-0.5	1.5	(-0.5)
-2.0	0.5	2.0	(0.5)
(-4.5)	(2.5)	(4.5)	

12.

0.5	2.0	1.0	(3.5)
-1.5	-0.5	2.5	(0.5)
-2.0	1.5	-1.0	(-1.5)
(-3.0)	(3.0)	(2.5)	

Answers

13.

-1.0	-1.5	2.0	(-0.5)
2.5	1.0	0.5	(4.0)
-0.5	1.5	-2.0	(-1.0)
(1.0)	(1.0)	(0.5)	

14.

-2.0	-1.0	-1.5	(-4.5)
-0.5	2.5	0.5	(2.5)
1.5	1.0	2.0	(4.5)
(-1.0)	(2.5)	(1.0)	

15.

-2.0	2.0	-0.5	(-0.5)
1.5	0.5	1.0	(3.0)
-1.0	2.5	-1.5	(0)
(-1.5)	(5.0)	(-1.0)	

16.

0.5	-2.0	1.0	(-0.5)
-1.5	1.5	-1.0	(-1.0)
2.0	-0.5	2.5	(4.0)
(1.0)	(-1.0)	(2.5)	

17.

-0.5	-1.0	1.0	(-0.5)
2.5	-2.0	2.0	(2.5)
1.5	-1.5	0.5	(0.5)
(3.5)	(-4.5)	(3.5)	

18.

-0.5	-1.0	2.0	(0.5)
0.5	-1.5	-2.0	(-3.0)
1.0	1.5	2.5	(5.0)
(1.0)	(-1.0)	(2.5)	

Answers

19.

0.5	-1.0	-1.5	(-2.0)
-2.0	1.5	2.0	(1.5)
1.0	2.5	-0.5	(3.0)
(-0.5)	(3.0)	(0)	

20.

-2.0	1.0	-1.0	(-2.0)
2.5	1.5	0.5	(4.5)
2.0	-1.5	0.5	(0)
(2.5)	(1.0)	(-1.0)	

21.

1.0	0.5	2.5	(4.0)
-2.0	1.5	-1.5	(-2.0)
-0.5	-1.0	2.0	(0.5)
(-1.5)	(1.0)	(3.0)	

22.

2.5	-2.0	2.0	(2.5)
-0.5	0.5	-1.5	(-1.5)
1.0	-1.0	1.5	(1.5)
(3.0)	(-2.5)	(2.0)	

23.

1.0	-2.0	0.5	(-0.5)
1.5	-0.5	-1.5	(-0.5)
2.0	-1.0	2.5	(3.5)
(4.5)	(-3.5)	(1.5)	

24.

1.5	-2.0	0.5	(0)
-1.0	-0.5	-1.5	(-3.0)
2.5	2.0	1.0	(5.5)
(3.0)	(-0.5)	(0)	

Answers

25.

0.5	-0.5	1.0	(1.0)
-1.0	-1.5	-2.0	(-4.5)
1.5	2.5	2.0	(6.0)
(1.0)	(0.5)	(1.0)	

26.

-2.0	-0.5	-1.0	(-3.5)
1.5	2.5	2.0	(6.0)
0.5	-1.5	1.0	(0)
(0)	(0.5)	(2.0)	

27.

1.5	1.0	-1.5	(1.0)
2.0	-1.0	-2.0	(-1.0)
-0.5	0.5	2.5	(2.5)
(3.0)	(0.5)	(-1.0)	

28.

-0.5	2.0	-1.5	(0)
0.5	2.5	-2.0	(1.0)
1.0	1.5	-1.0	(1.5)
(1.0)	(6.0)	(-4.5)	

29.

-2.0	-1.5	-1.0	(-4.5)
-0.5	0.5	1.0	(1.0)
1.5	0.5	2.5	(6.0)
(-1.0)	(1.0)	(2.5)	

30.

-1.5	-2.0	2.5	(-1.0)
-1.0	1.5	-0.5	(0)
2.0	0.5	1.0	(3.5)
(-0.5)	(0)	(3.0)	

PRIME Center

Answers

31.

2.5	2.0	1.0	(5.5)
-1.5	0.5	-2.0	(-3.0)
1.5	-0.5	-1.0	(0)
(2.5)	(2.0)	(-2.0)	

32.

-2.0	-1.5	-1.0	(-4.5)
-0.5	0.5	1.0	(1.0)
1.5	2.0	2.5	(6.0)
(-1.0)	(1.0)	(2.5)	

33.

-0.5	1.0	-1.5	(-1.0)
2.5	-2.0	2.0	(2.5)
-1.0	1.5	0.5	(1.0)
(1.0)	(0.5)	(1.0)	

34.

1.0	-1.5	0.5	(0)
2.0	1.5	-2.0	(1.5)
-0.5	2.5	-1.0	(1.0)
(2.5)	(2.5)	(-2.5)	

35.

1.0	-1.5	0.5	(0)
-2.0	1.5	2.0	(1.5)
2.5	-0.5	-1.0	(1.0)
(1.5)	(-0.5)	(1.5)	

36.

1.0	-1.0	2.5	(2.5)
-1.5	2.0	-0.5	(0)
0.5	-2.0	1.5	(0)
(0)	(-1.0)	(3.5)	

PRIME Center

Answers

37.

-2.0	1.5	-1.5	(-2.0)
1.0	0.5	-0.5	(1.0)
-1.0	2.5	2.0	(3.5)
(-2.0)	(4.5)	(0)	

38.

-0.5	2.0	1.0	(2.5)
-1.5	-1.0	0.5	(-2.0)
1.5	-2.0	2.5	(2.0)
(-0.5)	(-1.0)	(4.0)	

39.

-1.0	0.5	1.0	(0.5)
-2.0	2.5	-1.5	(-1.0)
-0.5	2.0	1.5	(3.0)
(-3.5)	(5.0)	(1.0)	

40.

0.5	1.5	2.5	(4.5)
-1.0	1.0	2.0	(2.0)
-0.5	-1.5	-2.0	(-4.0)
(-1.0)	(1.0)	(2.5)	

41.

-2.0	2.0	-0.5	(-0.5)
-1.5	-1.0	1.0	(-1.5)
2.5	1.5	0.5	(4.5)
(-1.0)	(2.5)	(1.0)	

42.

2.5	-2.0	0.5	(1.0)
-1.5	-1.0	1.5	(-1.0)
-0.5	1.0	2.0	(2.5)
(0.5)	(-2.0)	(4.0)	

PRIME Center   MATHadazzles Volume 8

Answers

43.

-1.5	1.0	2.5	(2.0)
0.5	-1.0	-2.0	(-2.5)
-0.5	2.0	1.5	(3.0)
(-1.5)	(2.0)	(2.0)	

44.

-2.0	-0.5	2.5	(0)
2.0	-1.0	0.5	(1.5)
1.5	1.0	-1.5	(1.0)
(1.5)	(-0.5)	(1.5)	

45.

-2.0	-1.5	-1.0	(4.5)
-0.5	0.5	1.0	(1.0)
1.5	2.0	2.5	(6.0)
(-1.0)	(1.0)	(2.5)	

46.

1.0	0.5	-1.0	(0.5)
-1.5	2.5	-0.5	(0.5)
2.0	1.5	-2.0	(1.5)
(1.5)	(4.5)	(-3.5)	

47.

-2.0	1.0	-1.5	(-2.5)
2.5	2.0	1.5	(6.0)
-1.0	0.5	-0.5	(-1.0)
(-0.5)	(3.5)	(-0.5)	

48.

1.0	2.0	-1.0	(2.0)
-1.5	1.5	-2.0	(-2.0)
2.5	0.5	-0.5	(2.5)
(2.0)	(4.0)	(-3.5)	

PRIME Center   MATHadazzles Volume 8

Answers

49.

0.5	-1.0	2.0	(1.5)
-2.0	1.0	1.5	(0.5)
-1.5	-0.5	2.5	(0.5)
(-3.0)	(-0.5)	(6.0)	

50.

2.0	1.5	-1.0	(2.5)
2.5	-1.5	-0.5	(0.5)
0.5	1.0	-2.0	(-0.5)
(5.0)	(1.0)	(-3.5)	

51.

2.0	2.5	-1.5	(3.0)
1.5	-2.0	-0.5	(-1.0)
-1.0	1.0	0.5	(0.5)
(2.5)	(1.5)	(-1.5)	

52.

-2.0	1.5	-1.0	(-1.5)
1.0	-1.5	2.5	(2.0)
-0.5	2.0	0.5	(2.0)
(-1.5)	(2.0)	(2.0)	

53.

0.5	-0.5	2.0	(2.0)
1.5	-1.5	1.0	(1.0)
2.5	-2.0	-1.0	(-0.5)
(4.5)	(-4.0)	(2.0)	

54.

2.5	2.0	1.5	(6.0)
1.0	0.5	-0.5	(1.0)
-1.0	-1.5	-2.0	(-4.5)
(2.5)	(1.0)	(-1.0)	

PRIME Center   MATHadazzles Volume 8

Answers

55.

2.5	2.0	-2.0	(2.5)
1.0	-1.0	1.5	(1.5)
-1.5	0.5	-0.5	(-1.5)
(2.0)	(1.5)	(-1.0)	

56.

1.5	0.5	-1.5	(0.5)
1.0	2.0	-1.0	(2.0)
-2.0	2.5	-0.5	(0)
(0.5)	(5.0)	(-3.0)	

57.

0.5	-0.5	2.5	(2.5)
1.5	-1.5	2.0	(2.0)
1.0	-1.0	-2.0	(-2.0)
(3.0)	(-3.0)	(2.5)	

58.

-0.5	1.0	-1.0	(-0.5)
1.5	-2.0	0.5	(0)
2.5	-1.5	2.0	(3.0)
(3.5)	(-2.5)	(1.5)	

59.

-2.0	-0.5	1.5	(-1.0)
-1.5	2.5	1.0	(2.0)
-1.0	2.0	0.5	(1.5)
(-4.5)	(4.0)	(3.0)	

60.

2.0	-1.5	1.0	(1.5)
2.5	-2.0	-0.5	(0)
1.5	0.5	-1.0	(1.0)
(6.0)	(-3.0)	(-0.5)	

PRIME Center   MATHadazzles Volume 8

Answers

61.

-2.0	1.5	1.0	(0.5)
2.0	-0.5	-1.5	(0)
0.5	-1.0	2.5	(2.0)

(0.5) (0) (2.0)

62.

-2.0	-0.5	2.0	(-0.5)
1.0	2.5	-1.0	(2.5)
0.5	-1.5	1.5	(0.5)

(-0.5) (0.5) (2.5)

63.

-2.0	-1.0	1.0	(-2.0)
1.5	-1.5	2.0	(2.0)
-0.5	2.5	0.5	(2.5)

(-1.0) (0) (3.5)

64.

1.5	2.0	2.5	(6.0)
-2.0	1.0	-0.5	(-1.5)
-1.0	0.5	-1.5	(-2.0)

(-1.5) (3.5) (0.5)

65.

2.0	0.5	-1.5	(1.0)
1.5	-0.5	-2.0	(-1.0)
1.0	-1.0	2.5	(2.5)

(4.5) (-1.0) (-1.0)

66.

-2.0	-0.5	1.5	(-1.0)
2.0	0.5	-1.5	(1.0)
-1.0	1.0	2.5	(2.5)

(-1.0) (1.0) (2.5)

PRIME Center   MATHadazzles Volume 8

Answers

67.

1.5	2.0	-1.0	(2.5)
1.0	-1.5	0.5	(0)
-2.0	-0.5	2.5	(0)
(0.5)	(0)	(2.0)	

68.

-2.0	-0.5	1.5	(-1.0)
-1.5	0.5	2.0	(1.0)
-1.0	1.0	2.5	(2.5)
(-4.5)	(1.0)	(6.0)	

69.

1.0	2.0	0.5	(3.5)
-1.5	-2.0	-1.0	(-4.5)
-0.5	2.5	1.5	(3.5)
(-1.0)	(2.5)	(1.0)	

70.

2.0	2.5	-0.5	(4.0)
0.5	-1.5	1.0	(0)
-1.0	-2.0	1.5	(-1.5)
(1.5)	(-1.0)	(2.0)	

71.

-2.0	-1.5	2.0	(-1.5)
1.0	1.5	0.5	(3.0)
2.5	-1.0	-0.5	(1.0)
(1.5)	(-1.0)	(2.0)	

72.

-0.5	-1.0	0.5	(-1.0)
-1.5	1.5	2.5	(2.5)
1.0	2.0	-2.0	(1.0)
(-1.0)	(2.5)	(1.0)	

Answers

73.

2.0	-2.0	2.5	(2.5)
-1.0	1.5	-1.5	(-1.0)
1.0	-0.5	0.5	(1.0)

(2.0) (-1.0) (1.5)

74.

-1.0	-1.5	1.0	(-1.5)
-2.0	2.5	2.0	(2.5)
0.5	1.5	-0.5	(1.5)

(-2.5) (2.5) (2.5)

75.

2.5	-1.0	-2.0	(-0.5)
-0.5	2.0	0.5	(2.0)
-1.5	1.5	1.0	(1.0)

(0.5) (2.5) (-0.5)

76.

0.5	-2.0	1.0	(-0.5)
-0.5	2.5	-1.0	(1.0)
-1.5	2.0	1.5	(2.0)

(-1.5) (2.5) (1.5)

77.

-1.5	-2.0	1.0	(-2.5)
1.5	2.0	0.5	(4.0)
2.5	-0.5	-1.0	(1.0)

(2.5) (-0.5) (0.5)

78.

0.5	-0.5	1.0	(1.0)
-2.0	2.5	-1.0	(-0.5)
2.0	-1.5	1.5	(2.0)

(0.5) (0.5) (1.5)

The *MATHadazzles, Mind Stretch Puzzles* series includes:

- **Volume 1 Reasoning with Numbers**
- **Volume 2 Reasoning with Whole Numbers**
- **Volume 3 Reasoning with Integers**
- **Volume 4 Reasoning with Fractions**
- **Volume 5 Reasoning with Decimals**
- **Volume 6 Reasoning Algebraically**
- **Volume 7 Reasoning Algebraically**
- **Volume 8 Reasoning Algebraically with Decimals**

Contributors to Volumes 1, 2, and 3 are middle school teachers in the Greater Phoenix area.

Contributors to Volumes 4 and 5 are middle school students from the Greater Phoenix area.

Contributers to Volumes 6, 7 and 8 are high school students from the Greater Phoenix area who participated in the NSF Project AMP (#1509105).

www.ingramcontent.com/pod-product-compliance
Lightning Source LLC
Chambersburg PA
CBHW070103210526
45170CB00012B/723